SHOULD SPACEX AND NASA COLONIZE VENUS INSTEAD OF MARS?
Why Humanity's Future May Lie in Its Skies

A Deep Dive into the Science, Strategy, and Vision for Colonizing Our Volcanic Twin

Tommy S. Manley

Copyright ©Tommy S. Manley, 2024.

All rights reserved. No part of this publication may be reproduced, distributed, or transmitted in any form or by any means, including photocopying, recording, or other electronic or mechanical methods, without the prior written permission of the publisher, except in the case of brief quotations embodied in critical reviews and certain other noncommercial uses permitted by copyright law.

Table of Contents

Introduction..3

Chapter 1: Revisiting Venus — More Than a 'Hot' Planet..6

Chapter 2: The Science of Venus — Earth's Volcanic Twin...14

Chapter 3: The Challenges of Venus Colonization..22

Chapter 4: The Concept of Cloud Cities — Humanity's Best Hope..31

Chapter 5: Comparing Mars and Venus as Habitats for Humanity..39

Chapter 6: Terraforming Venus — Dreams, Science, and Future Vision..49

Chapter 7: The Current Roadmap — Upcoming Venus Missions and Goals..59

Chapter 8: The Role of Private Space Exploration in Venus Colonization..67

Chapter 9: The Future of Multiplanetary Civilization..76

Conclusion...84

Introduction

Mars has captured the imagination of millions as humanity's next potential home, drawing in explorers, scientists, and dreamers alike who envision a red frontier filled with possibilities. Mars, with its valleys, canyons, and polar ice caps, offers a rugged allure—a new frontier on which to establish a human presence. With its relatively mild conditions in comparison to other planets, Mars feels almost within reach, a stepping stone in our solar system that could mark humanity's first step toward living beyond Earth. But as we look to Mars, are we overlooking a closer, equally intriguing neighbor?

Consider Venus, our nearest planetary companion and an extraordinary candidate for human exploration. Venus, often dismissed as too hostile and unpredictable, holds secrets that could reshape our perspective on what makes a planet habitable. Known as Earth's "volcanic twin" due to its similar size, gravity, and geological features, Venus shares surprising characteristics with our home planet. Despite its scorching surface and dense, toxic atmosphere, Venus has layers of intrigue—both literal and scientific—that make it far more than just a "hellish" world.

High above its turbulent surface, in the temperate layers of its atmosphere, lies a zone where conditions resemble Earth's in both pressure and temperature. In these atmospheric heights, the potential for human exploration, and perhaps even floating colonies, is real. Unlike Mars, where humanity would grapple with harsh dust storms, thin air, and freezing temperatures, Venus offers a surprising alternative. A world where human habitats might float among acid clouds, held aloft in the gentle pull of its atmosphere, closer in some ways to Earth than we may realize.

The purpose of this book is to re-examine Venus as a viable destination for humanity's future, challenging the assumption that Mars should be our primary target. Through a blend of scientific analysis, historical context, and future vision, we will embark on a journey to understand why Venus deserves a second look. By exploring its complex atmosphere, studying the remarkable landscapes hidden beneath its dense clouds, and considering the innovative ideas for human settlement, this journey will unveil a world of possibility. Venus, with its mysteries and challenges, may well be humanity's next frontier—not below its fiery surface, but above, where the skies offer a glimpse of a new kind of life beyond Earth.

Chapter 1: Revisiting Venus — More Than a 'Hot' Planet

In the realm of planetary exploration, Venus has long held a reputation as the "planet of fire and fury," an image shaped by its relentless extremes and near-inhospitable atmosphere. To most, Venus is a world of hostile conditions: temperatures high enough to melt lead, clouds thick with sulfuric acid, and an air pressure akin to being submerged under half a mile of ocean. For centuries, these conditions have made Venus seem more like a vision of purgatory than a potential second home for humanity, casting it as a cautionary tale rather than a destination of hope. Ancient civilizations saw Venus as a bright, mysterious figure in the sky, appearing just before sunrise or at sunset—an enigmatic companion to Earth, embodying both allure and danger. This "morning star" later turned out to be a planet shrouded in clouds that conceal an intensely volatile landscape, forever altering its allure.

Yet, beneath these forbidding layers lies a complex world with much more to offer than meets the eye. Venus occupies a unique place in our solar system, positioned as the second planet from the Sun and, unlike most of its planetary neighbors, orbiting closer to Earth than any other planet. This proximity makes it not only visible but prominent in our night sky. It's almost as if Venus has been hiding in plain sight, a close neighbor we have yet to truly understand. Nearly the same size as Earth, with a diameter only slightly smaller, Venus has often been referred to as Earth's "sister planet." This resemblance goes beyond size alone; its gravity is similar to Earth's, and it even boasts a thick, multi-layered atmosphere, although one saturated with carbon dioxide.

The Venusian surface, hidden under dense cloud cover, is a land of endless volcanic plains and towering mountains. We know it hosts over a thousand major volcanoes and volcanic features, hinting at a history of intense geological activity.

But unlike Earth's tectonic plates, Venus's crust is solid and immobile, with its volcanic flows likely shaping the landscape in bursts of fiery intensity rather than steady shifts. This unique tectonic makeup has led scientists to believe that Venus underwent massive resurfacing events, possibly due to catastrophic volcanic activity, which erased much of its ancient history. It's a planet that, while eerily Earth-like in some ways, seems to have taken a dramatic turn in a different direction.

In terms of rotation, Venus is also curiously unique. Unlike most planets in the solar system, Venus rotates slowly and in the opposite direction to its orbit around the Sun. This retrograde rotation, combined with its slow spin, means that a day on Venus lasts longer than a year. Moreover, with a nearly upright axial tilt, Venus lacks the familiar seasonal cycles we experience on Earth. This gives Venus a static, perpetual state that feels as if time itself is distorted on this fiery planet.

Venus's magnetic field, or lack thereof, further distinguishes it. Without a protective magnetic barrier like Earth's, Venus is vulnerable to the Sun's intense radiation, which directly interacts with its upper atmosphere. This exposure likely contributes to its harsh atmospheric conditions, with solar winds blowing particles away and gradually altering its atmospheric makeup over millions of years.

Through the lens of science, Venus appears as a beautifully complex world. It's a planet sculpted by volcanic force, governed by strange rhythms of time, and draped in a thick veil of clouds that act as both shield and prison. The scientific mysteries of Venus are as captivating as its extremes, revealing a world that is neither wholly alien nor entirely familiar—a strange and volatile twin of Earth, waiting to be uncovered.

Despite Venus's proximity and intriguing similarities to Earth, our exploration of this planet has been surprisingly limited. This neglect isn't due to a lack of interest or scientific value; rather, it stems from the extraordinary challenges Venus presents and the harsh reality of what it takes to study it up close. In the early days of space exploration, Venus initially seemed like a prime candidate for discovery. Its size and position in the solar system sparked curiosity, and for a time, scientists even speculated that its thick clouds might conceal lush, tropical landscapes or oceans. Yet as missions began probing its atmosphere and gathering data, the dream of a potentially habitable Venus faded into something far more daunting.

The first missions sent toward Venus were met with relentless hostility. In the 1960s and 70s, when Soviet probes from the Venera program attempted to reach its surface, they faced crushing atmospheric pressure, searing heat, and clouds laden with sulfuric acid. Early probes were destroyed long before they could send meaningful data back to Earth. Those that did survive, thanks to ever-stronger engineering, managed only a few minutes of transmission before the punishing environment claimed them as well. Unlike Mars, which offers a cold but relatively stable landscape, Venus defies conventional exploration methods, and every mission sent there faces a battle for survival.

With each new failure, Venus's reputation as an unforgiving world grew, and so did the costs and risks associated with studying it. As technology advanced, however, Mars emerged as a more feasible target, offering a challenging but comparatively accessible environment. With its thin atmosphere, predictable weather, and surface conditions that, while harsh, allowed for rovers to land and operate for extended periods, Mars became a natural focus. The allure of a surface that could support robotic exploration—and perhaps one day, human colonies—only added to its appeal, drawing resources, research, and ambition away from Venus.

Moreover, Mars offered the tantalizing possibility of water, and with it, a chance of life. This potential gave Mars a uniquely compelling edge, especially as scientists and the public alike became increasingly captivated by the search for extraterrestrial life. Resources poured into Martian exploration, from orbiters and landers to the iconic rovers, each new mission promising to reveal Mars's secrets while Venus languished in the shadows.

The more we discovered about Venus, the clearer it became that the traditional methods of planetary exploration were largely ineffective against its extreme conditions. Its high-pressure atmosphere is dense enough to crush most standard equipment, and its clouds, filled with corrosive sulfuric acid, eat away at electronics and surfaces. Add to this the blistering heat of its surface, hotter than even Mercury, and it's easy to see why Venus gradually slipped down the priority list.

This redirection of focus didn't mean Venus was without value or intrigue, but Mars, with its more manageable conditions and potential for habitability, offered a lower barrier to entry. The fascination with Mars has only grown over the years, as we dream of human missions and eventual colonization. Meanwhile, Venus remains shrouded in mystery, its unique and challenging environment demanding new, inventive approaches to exploration.

Chapter 2: The Science of Venus — Earth's Volcanic Twin

Venus's atmosphere is like an intense and exaggerated version of Earth's greenhouse effect, presenting a world shrouded in thick, carbon dioxide-heavy clouds that envelop the entire planet. Unlike Earth, where CO_2 forms a small fraction of our atmosphere, Venus's air is dominated by carbon dioxide, accounting for over 96% of its atmospheric makeup. This dense CO_2 layer traps heat in a runaway greenhouse effect, creating surface temperatures that soar to 460°C (860°F), hotter even than Mercury, despite Venus being farther from the Sun. Layers of sulfuric acid clouds hover above, adding a uniquely hostile element, forming a permanent blanket that blocks direct sunlight while trapping heat with relentless efficiency.

High above the surface, these sulfuric acid clouds create an environment so corrosive that they can erode any material over time, posing extreme challenges for any equipment or probe attempting to study Venus up close. Yet, within these very clouds lies a zone where pressure and temperature are less severe, hinting at a space where exploration, if not habitability, might be possible. However, closer to the surface, the atmospheric pressure becomes an unyielding force. At approximately 92 times the pressure on Earth, it's comparable to the crushing conditions half a mile underwater on Earth, a challenge that has made even the most robust space probes struggle for survival.

In addition to its extreme atmospheric composition, Venus also exhibits fascinating planetary characteristics that set it apart from Earth. Unlike most planets, Venus rotates in the opposite direction of its orbit, meaning that if you stood on its surface, you'd see the Sun rise in the west and set in the east. This retrograde rotation is exceptionally slow, taking 243 Earth days to complete a single rotation. Surprisingly, this also means a day on Venus—measured from one sunrise to the next—is longer than its year, as it takes only 225 days for Venus to orbit the Sun. This unusual cycle has profound effects on the planet's climate, creating a strange temporal landscape where day and night stretch across Earth months.

Adding to its uniqueness, Venus has an almost upright axis, so it doesn't experience the seasonal changes familiar to Earth. Its lack of a magnetic field further leaves it vulnerable to solar winds, which continually strip away particles from its upper atmosphere and expose the planet to intense radiation. These winds and radiation likely play a role in maintaining the strange, high-speed super-rotation of Venus's upper atmosphere, where winds can reach speeds of over 360 kilometers per hour (224 miles per hour), circling the entire planet in just a few Earth days.

With its crushing pressure, scorching temperatures, and acidic clouds, Venus feels like an amplified version of Earth, embodying the very extremes of planetary conditions. And yet, these harsh characteristics make Venus scientifically valuable, offering insights into climate dynamics, greenhouse effects, and atmospheric chemistry on a scale we can't replicate on Earth. It's a planet that mirrors some of Earth's foundational qualities, albeit in a warped, hostile fashion, revealing how similar starting conditions can diverge into radically different worlds.

Venus's surface, though hidden beneath thick clouds, reveals an array of geological features that bear a striking resemblance to Earth's. Its landscape is sculpted with vast volcanic plains, mountainous regions, and deep valleys that hint at a dynamic geologic past. Venus has thousands of volcanoes dotting its surface, some of which are as large as Earth's most massive volcanic formations. These volcanoes, however, don't fit into the same tectonic narrative as those on Earth, where the shifting of plates creates mountains, faults, and oceanic trenches. Venus's crust is a single, unbroken layer with no tectonic plates, meaning that volcanic activity likely occurred through the planet's own internal pressure releasing in massive, planet-altering eruptions. This creates a unique scientific scenario where Venus appears almost frozen in a moment of its geological evolution, potentially preserving evidence of past volcanic activity on a planetary scale.

In addition to its volcanic landscape, Venus boasts mountains that tower above its plains, like the towering Maxwell Montes, which rivals Everest in height. These mountains, along with ridges and valleys, are scattered across Venus's surface, forming a landscape that mirrors many of Earth's natural formations. Yet, the forces shaping them are quite different, with Venus's solid crust restricting movement and allowing volcanic flows to become major landscape-shaping events. The discovery of such Earth-like features on Venus presents a fascinating opportunity for scientists: studying them could offer a glimpse into how Earth might look if it were subjected to a similar atmospheric and geological environment, or even how planets evolve without tectonic movement. Venus provides an alternate version of geologic evolution, a snapshot of what Earth's terrain might resemble under extreme atmospheric conditions.

Adding to its unusual nature are Venus's atmospheric phenomena, where layers of supercritical carbon dioxide dominate. In this state, carbon dioxide behaves neither like a gas nor a liquid but rather as a "supercritical fluid," a condition brought on by the planet's immense pressure and temperature. This supercritical layer gives Venus's lower atmosphere an almost fluid-like density, further distinguishing it from Earth's atmosphere. In this supercritical state, CO_2 behaves more like a liquid than a gas, enveloping the surface in a thick, soupy medium that transmits heat efficiently, preventing cooling at the surface.

Above this dense layer, clouds of sulfuric acid swirl in a perpetual, otherworldly sky, their composition far different from Earth's water vapor clouds. While sulfuric acid rain does form within these clouds, it never reaches the surface; the intense heat of the lower atmosphere causes the acid droplets to evaporate before they can descend fully, creating an environment of "acid rain" that perpetually hangs in the Venusian skies without ever touching the ground.

One of the most intriguing aspects of Venus's atmosphere is its super-rotational wind pattern, where the upper atmosphere whips around the planet at breakneck speeds, circling it in as little as four Earth days. These high-altitude winds travel up to 60 times faster than Venus's own rotation, creating a constant westward flow that's far stronger than anything seen on Earth. This rapid super-rotation remains somewhat mysterious, with scientists still investigating how such speeds are sustained despite Venus's slow rotational period. The winds create a powerful jet stream that carries the thick clouds in a continuous swirl, giving Venus an atmospheric dynamism that contrasts starkly with its still, unmoving surface below.

These extraordinary atmospheric phenomena transform Venus into a unique laboratory for understanding planetary climates and atmospheres. By studying Venus's supercritical CO_2 layer, sulfuric acid clouds, and high-speed winds, scientists gain valuable insights into extreme atmospheric dynamics and greenhouse effects. The extreme contrasts and similarities between Venus and Earth provide a vivid reminder of how planetary conditions can diverge in dramatic ways, underscoring the delicate balance that makes Earth habitable and highlighting the science that might help us preserve it.

Chapter 3: The Challenges of Venus Colonization

The surface of Venus presents an almost insurmountable challenge for exploration. Its blistering heat, crushing pressure, and toxic atmosphere combine to create an environment that is nearly as hostile as one could imagine. At a scorching 460°C (860°F), the Venusian surface temperature is hot enough to melt lead. The heat is so intense that it creates a kind of feedback loop, where the ground absorbs and radiates heat in all directions, trapped by the thick, carbon dioxide-heavy atmosphere. This extreme greenhouse effect keeps the planet's surface at a constant, searing temperature, day or night, with little seasonal variation.

Adding to the hellish environment is the atmospheric pressure, which is about 92 times that of Earth's at sea level—equivalent to the crushing force felt a half mile underwater. This pressure would make the experience of standing on Venus's surface feel like being inside an industrial-grade press, capable of collapsing most structures and devices. Even if a machine could withstand the heat, the pressure would remain a relentless force, straining every inch of its design. Venus's atmosphere, filled with toxic clouds of sulfuric acid, poses yet another obstacle. Not only is it inhospitable to any form of life as we know it, but the corrosive acid in the clouds erodes anything that tries to endure within it, breaking down machinery over time and rendering typical materials useless.

These extreme conditions have thwarted many exploration efforts in the past. The Soviet Union's Venera program, which sent a series of probes to Venus in the 1960s and 70s, made some of the earliest attempts to pierce the Venusian clouds and reach the surface. Yet, each probe encountered a brutal environment that tested the very limits of technology at the time. The first few Venera probes succumbed almost instantly; crushed, melted, or simply obliterated by the unforgiving conditions. Those that made it to the surface survived only moments before their instruments were overwhelmed. Early probes, designed to withstand harsh environments, were not yet equipped for the level of resistance Venus demanded, and any parachutes used for landing quickly failed under the weight of the thick atmosphere.

As the Venera program advanced, engineers developed stronger, more resilient designs, fortifying probes with reinforced casings and eliminating weak points wherever possible. By the time of Venera 13 and 14, which were the most successful missions, the probes managed to survive for a few hours, transmitting invaluable data and the first images of Venus's rocky, barren surface before succumbing to the inevitable breakdown from the extreme conditions. These fleeting windows of success revealed some of Venus's secrets, but they also underscored how limited our technology was in penetrating this hostile world.

While technology has progressed since then, Venus remains a formidable opponent. Even today, the extreme heat, corrosive atmosphere, and intense pressure require innovations that go beyond conventional engineering. To return to Venus, any lander or probe would need to incorporate materials capable of withstanding acid erosion, high temperatures, and immense pressure—all while maintaining functionality long enough to capture data. Developing such technology comes at a high cost and requires complex, precise engineering, which has pushed Venus exploration further down the priority list in favor of easier, more accessible planets.

Despite these hurdles, Venus remains a tantalizing scientific target, a world that challenges the boundaries of our technology and expertise. Each new attempt at exploration, while fraught with risk, holds the promise of revealing clues about our own planet's climate, atmospheric dynamics, and geological processes. Venus tests not only the durability of our machines but the limits of human curiosity, daring us to push further into the unknown, despite the daunting barriers it poses.

The Soviet Union's Venera program stands as one of the most remarkable and ambitious attempts to study Venus, pushing the limits of engineering and space exploration in an era when the technology needed to survive the planet's harsh conditions was still in its infancy. Beginning in the 1960s and spanning nearly two decades, the Venera missions provided humanity's first glimpses beneath Venus's clouded exterior, delivering invaluable insights into a world as hostile as it was mysterious. Though many early attempts were thwarted by the planet's intense pressure and heat, each mission brought engineers closer to understanding what it would take to explore Venus.

The initial Venera missions, launched between 1961 and 1969, were met with sobering challenges. Venus's extreme conditions crushed and melted the first probes long before they reached the surface, revealing just how hostile the planet truly was. These early failures were vital learning experiences, highlighting the need for robust designs and better protective measures. The engineers quickly learned that a typical probe, designed to explore planetary bodies with more forgiving environments, was ill-suited for Venus. With each new mission, they refined the designs, reinforcing the structures, improving heat resistance, and adjusting for the incredible atmospheric density.

It was the later missions, from Venera 7 onwards, that marked significant milestones. Venera 7, launched in 1970, became the first spacecraft to successfully land on another planet and transmit data back to Earth, albeit for just 23 minutes before succumbing to the conditions. Even this brief transmission provided valuable information, proving that probes could survive the descent through Venus's atmosphere and operate, albeit briefly, on its surface. This mission set the stage for a series of increasingly resilient probes that would pave the way for further exploration.

The achievements of Venera 9 and 10 were even more groundbreaking. In 1975, these missions not only landed on Venus but sent back the first images from its surface, showing a rocky, barren landscape devoid of visible life and covered in fragmented, sharp stones. The photos were a historical first, giving humanity its initial look at the Venusian surface. These images confirmed what scientists had suspected: Venus was an intensely hostile environment, with landscapes shaped by immense volcanic activity and atmospheric pressure that flattened the landscape in an otherworldly stillness.

Venera 13 and 14, launched in 1981, represented the height of the program's success. These probes carried advanced instruments for analyzing the planet's surface composition, atmospheric chemistry, and physical properties, managing to survive for just over two hours on the surface—an impressive feat given the conditions. They transmitted detailed data on the rocks and soil, confirming a volcanic composition similar in some ways to Earth's basalt formations, yet altered by the extreme conditions. These missions also provided the first audio recordings from another planet, capturing the eerie, muffled sounds of Venus's dense atmosphere.

Yet despite these achievements, the Venera program was limited by the technology of its time. The probes' lifespans were measured in minutes and hours, and each mission could only capture a small fraction of data before the hostile environment took its toll. Venus's crushing pressure, intense heat, and corrosive atmosphere ultimately claimed every probe, leaving scientists with tantalizing fragments of information but also many unanswered questions. Each Venera mission revealed as much about the limitations of human technology as it did about Venus itself, underscoring how formidable the planet's conditions truly are.

The Venera program's legacy is profound. It not only pioneered planetary exploration but also set a high bar for resilience and ingenuity in space engineering. These missions demonstrated that Venus, for all its hostility, was a world worth studying, sparking interest that continues to influence planetary science today. The insights from Venera laid the groundwork for modern Venus exploration, providing invaluable data that scientists and engineers still reference as they design the next generation of probes to brave this enigmatic world. In many ways, the Venera missions were a testament to human curiosity and determination, capturing fleeting glimpses of a planet that still guards its secrets fiercely.

Chapter 4: The Concept of Cloud Cities — Humanity's Best Hope

As scientists continue to grapple with the hostile surface of Venus, some have suggested a bold and inventive solution: floating cities high above the planet's surface. This concept, which sounds as if it were lifted from science fiction, is not only possible but uniquely viable on Venus. Instead of contending with the extreme heat, crushing pressure, and corrosive atmosphere at ground level, human habitats could hover in the planet's upper atmosphere, where conditions are surprisingly similar to those on Earth. In this "cloud city" scenario, large, buoyant structures would drift above Venus's infernal landscape, sustained by the very air that would otherwise be toxic on the surface.

The key to this idea lies in what scientists refer to as the "Goldilocks Zone" within Venus's atmosphere. Around 50 to 60 kilometers above the surface, the conditions align in a rare balance, offering a habitable pocket where temperatures range between 30 to 50°C (86 to 122°F) and atmospheric pressure is roughly equivalent to Earth's at sea level. At this altitude, Venus's atmosphere is still dense, but the pressure is manageable, creating an environment in which Earth-like materials and life-support systems could function without the need for extreme reinforcement.

Here, in this temperate layer, the challenges of Venusian exploration shift from being about survival to about endurance and innovation. Within the Goldilocks Zone, air pressure sits at a comfortable 1 bar, nearly identical to what we experience on Earth's surface, eliminating the need for heavy-duty pressurization. The temperature, though warm, is within a tolerable range, especially for advanced materials designed to withstand a slight increase in heat. In this region, the toxic clouds are still present but pose a manageable risk; sulfuric acid droplets can be mitigated with coatings like Teflon or other acid-resistant materials, allowing habitats to endure the conditions without rapid degradation.

Interestingly, the dense atmosphere of Venus at this altitude presents an advantage: with an abundance of carbon dioxide—much denser than Earth's air—a simple balloon filled with breathable air would naturally float. This density differential provides an opportunity to construct habitats that remain buoyant with minimal energy input, using the basic principle of lighter-than-air technology. These floating cities could be expansive and stable, offering both living quarters and research facilities for prolonged stays.

Living above the surface of Venus might seem like an ambitious or even fantastical notion, yet the science behind it is straightforward and achievable. In the Goldilocks Zone, the temperature and pressure are manageable, and the upper atmosphere provides a relatively gentle environment in which to conduct exploration and scientific study. Sheltered from the worst of Venus's hostility, floating cities could offer humanity a way to colonize the clouds, establishing a presence on a neighboring world without touching down on its dangerous surface. Here, suspended in the sky, Venus becomes not an impossible destination but an entirely new frontier—one where life could thrive amidst the clouds, redefining the limits of exploration.

Designing habitats to float in Venus's upper atmosphere involves rethinking traditional concepts of structure and stability. In this unique environment, the idea centers around creating air-filled habitats that function much like massive balloons, using Earth-like air to create natural buoyancy in Venus's dense, carbon dioxide-laden atmosphere. Unlike Earth, where the atmosphere is composed primarily of nitrogen and oxygen, Venus's thick atmosphere of carbon dioxide makes our breathable air, which is lighter, a natural lifting gas. This allows structures to float stably, creating a self-sustaining system that requires minimal energy to remain aloft.

The basic design for these habitats would resemble large, spherical or cylindrical "balloons" with inner sections partitioned for living quarters, laboratories, and equipment storage. For stability, these habitats would need to be constructed with materials strong enough to withstand minor impacts and vibrations, yet flexible enough to accommodate the movement and expansion of air within. Acid-resistant materials would also be critical. Coating the exterior with materials such as Teflon, which resists corrosion, or advanced polymers designed to repel acid droplets, would shield the structures from the sulfuric acid present in Venus's cloud layers. With these coatings, the outer shell would be able to endure Venus's corrosive conditions over long periods, allowing inhabitants to safely conduct scientific research and daily activities.

Inside the habitats, the air pressure would be controlled to match Earth's standard pressure, creating a comfortable, breathable environment. Since this altitude on Venus offers nearly Earth-like pressure, the outer shell wouldn't require extreme reinforcements, making the habitat construction both lighter and more feasible. By balancing internal and external pressures, these floating habitats would not only be stable but also adaptable, allowing for modular expansions that could gradually transform small outposts into entire cities in the clouds.

This concept has roots in both fiction and reality. The idea of a "Cloud City" first captured public imagination through science fiction, most notably in *Star Wars*, where the fictional planet Bespin housed a floating metropolis suspended in the sky. This imaginative portrayal hinted at the potential for cities in high-atmosphere zones, and now, scientific progress suggests that similar concepts could be achievable on Venus.

Beyond fiction, real-world experiments have already tested this concept. The Soviet Union, in its exploration of Venus during the 1980s, deployed balloon probes as part of the Vega program. These probes, equipped with acid-resistant coatings, successfully floated within Venus's cloud layers for extended periods, gathering valuable atmospheric data while enduring the hostile conditions. These balloon probes demonstrated that it was possible to create resilient, acid-resistant systems capable of sustained operation in Venus's upper atmosphere, validating some of the foundational ideas behind floating habitats.

Drawing on these lessons from both science fiction and real-world probes, we can envision air-filled habitats that could one day hover in the Goldilocks Zone of Venus, protected against the acid-laden clouds. These cloud habitats would embody both practicality and ingenuity, using basic principles of buoyancy, atmospheric chemistry, and acid-resistant materials to establish humanity's presence above Venus's surface. The idea of living among the clouds of Venus, once the realm of fantasy, now sits at the edge of possibility, showing how imagination, when combined with technology, can redefine the frontiers of exploration.

Chapter 5: Comparing Mars and Venus as Habitats for Humanity

When comparing Venus and Mars for human exploration, the feasibility of surface operations on each planet could not be more different. Mars, with its cold but relatively stable environment, offers a surface that, while barren and devoid of liquid water, provides a predictable and tolerable setting for robotic and, potentially, human activity. Its thin atmosphere, primarily composed of carbon dioxide, is far from breathable, but it does offer some protection against small meteor impacts and minor radiation. Mars's surface, littered with dust and rock, presents a familiar landscape—one where rovers have navigated and where astronauts could, theoretically, set up shelters with the help of pressurization and thermal insulation.

On the other hand, Venus's surface is a different story entirely. The crushing atmospheric pressure at ground level, about 92 times that on Earth, is comparable to the conditions found half a mile underwater. This pressure, combined with surface temperatures of around 460°C (860°F), makes Venus an extraordinarily hostile environment. Even the most advanced materials would struggle to survive the heat and pressure for long, and any equipment deployed on Venus's surface would have to endure conditions far beyond those on Mars. While Mars's surface environment poses challenges, they are of a different scale altogether from Venus's, where conditions would melt lead and crush even the toughest machinery within hours.

Atmospheric conditions and available resources further differentiate these two worlds when it comes to potential colonization. Mars's atmosphere is thin, with only about 1% of Earth's pressure, and primarily composed of carbon dioxide. Although it lacks breathable oxygen, Mars's atmospheric carbon dioxide could be used to produce oxygen through chemical processes, providing a pathway to generating life-supporting air. Additionally, water ice exists beneath the Martian surface and in polar caps, which could be mined and processed to support both drinking water and fuel production for long-term missions. Solar power is a viable energy source on Mars, as its distance from the Sun still allows solar panels to generate substantial energy, though less efficiently than on Earth.

In contrast, Venus's atmosphere is thick, dense, and highly toxic. Its atmosphere contains minimal oxygen and is composed of over 96% carbon dioxide, with clouds laden with sulfuric acid. Breathing air is completely absent, and the need to protect against acid corrosion and the relentless greenhouse effect means that long-term surface operations are infeasible. However, high up in Venus's atmosphere lies a different story. At altitudes around 50 kilometers, a more Earth-like pressure and temperature emerge, creating a habitable zone where human colonies could theoretically exist. Here, the abundance of carbon dioxide could be utilized as a resource, converted to breathable oxygen with the right technology, while the dense atmosphere allows for the buoyant floating of habitats. Nevertheless, resource availability, such as water, remains a challenge. Unlike Mars, Venus offers little in terms of accessible water; any colony in Venus's atmosphere would need to rely on advanced extraction or transport systems, or synthetic water production processes, which could add logistical complexity.

In terms of sustainability, Mars holds an edge with its relatively stable surface environment, potential water resources, and manageable atmospheric conditions, allowing for easier construction and long-term planning. Venus, by contrast, presents a tantalizing possibility for atmospheric colonies, floating above its dangerous surface. While it lacks some of the accessible resources Mars offers, the floating colonies would be less vulnerable to meteor impacts and could tap into the dense atmosphere for energy and resource harvesting. Both planets, therefore, offer unique pathways for potential human settlement, with Mars presenting a more straightforward, surface-based approach and Venus challenging humanity to explore new architectural and engineering frontiers in the clouds. Each planet, with its distinct qualities, invites us to imagine and innovate differently, challenging our notions of what it means to extend human life beyond Earth.

When it comes to the logistical and economic aspects of exploring and potentially colonizing Mars and Venus, both planets present unique challenges and opportunities. Mars, often viewed as the "next frontier," is relatively accessible by current standards. Though still expensive, Mars missions have become more feasible due to recent advances in rocketry, particularly with companies like SpaceX developing reusable launch systems aimed at reducing costs. The flight time to Mars, while variable, typically ranges from six to nine months, allowing for the transport of humans and equipment within a manageable time frame. Given that Mars missions don't require groundbreaking new technology for orbit insertion, landing, or rover deployment, the cost structure, while substantial, is relatively predictable.

Venus, on the other hand, poses different logistical hurdles. Reaching Venus requires a similar flight duration to Mars, but the extreme conditions complicate any surface mission, making lander designs incredibly expensive due to the high demand for protective technology against the intense heat, pressure, and acid in the atmosphere. Missions to Venus's atmospheric layers, where floating habitats could potentially be deployed, require specialized high-altitude craft and acid-resistant materials, both of which push up costs. While Venus is closer to Earth than Mars, the need for sophisticated materials and technology offsets this advantage, making Venus missions significantly more complex from an engineering and financial standpoint.

In terms of long-term viability, Mars currently holds a cost advantage. Mars offers a stable surface environment where resources like ice can be mined and converted into water, oxygen, and even fuel. This potential for in-situ resource utilization (ISRU) reduces the need for supply shipments from Earth, which is essential for establishing a sustainable colony. On Mars, energy needs could be met with solar panels, and efforts are underway to develop nuclear power options for continuous power. Economically, Mars offers a clearer path to self-sufficiency, with infrastructure that could grow over time, turning initial investments into long-term assets.

Floating colonies in Venus's atmosphere, however, present a different model of long-term viability. While surface resources on Venus are largely inaccessible, the dense carbon dioxide in its upper atmosphere could be harvested and processed for fuel and life support, supporting some level of self-sustainability. The stability of a floating colony at approximately 50 kilometers in altitude offers a protected environment where solar power is plentiful, and habitats could be designed for low-energy maintenance, using natural buoyancy to stay afloat. Although more logistically complex and expensive to establish, a colony in Venus's atmosphere would operate in a protected environment, shielded from radiation and space debris, reducing certain maintenance and repair costs in the long run.

But beyond logistics and costs, the question of what's at stake is vital to understanding the potential of each planet for advancing human civilization. Colonizing Mars, with its stable surface, offers the chance to build a "second Earth," a backup for humanity where life could evolve to be partially self-sustaining. Mars could be a testing ground for technologies that support human life in hostile environments, from water recycling and closed-loop ecosystems to energy independence and building materials sourced from Martian regolith. If successful, Mars could lead to a permanent human foothold in space, serving as a platform for deeper exploration into the outer solar system.

Venus, though a less conventional choice, holds its own allure. Floating cities in the Venusian atmosphere would push the boundaries of architecture, engineering, and material science, driving technological advancements that could reshape how we design and build on Earth. A successful colony in Venus's atmosphere could reveal new ways to live sustainably and minimally, showcasing a model where self-contained habitats thrive in an otherwise uninhabitable environment. Additionally, Venus offers unparalleled opportunities for scientific study of climate dynamics, extreme atmospheric chemistry, and the greenhouse effect, insights that could be critical for understanding and addressing Earth's own environmental challenges.

Both Mars and Venus carry immense promise, but each represents a different vision for humanity's future in space. Mars offers a more straightforward path to expansion, providing a surface environment that, while difficult, is within our reach. Venus, on the other hand, offers the possibility of living in the skies, expanding human presence in an entirely novel way. Each planet has something vital to teach us about resilience, adaptation, and innovation, presenting humanity with not just a destination but a chance to redefine what it means to survive and thrive beyond Earth.

Chapter 6: Terraforming Venus — Dreams, Science, and Future Vision

Imagining a long-term vision for transforming Venus into a habitable world requires thinking on an ambitious, almost science-fiction scale. The concept of terraforming Venus, a planet whose surface conditions are far beyond what current technology can manage, presents a unique and complex challenge. Venus's thick, carbon dioxide-rich atmosphere traps immense heat, creating surface temperatures that make life as we know it impossible. However, the vision for a terraformed Venus involves gradually shifting its atmosphere and temperature to make the planet more Earth-like. While this may take hundreds or even thousands of years to achieve, scientists and visionaries have proposed bold strategies that could, theoretically, alter Venus in transformative ways.

One of the primary obstacles to terraforming Venus is its extreme heat, which is perpetuated by an intense greenhouse effect. To cool Venus down, one proposed solution is to reduce the amount of solar radiation that reaches the planet. A revolutionary way to accomplish this would be to install an array of sun-shading devices in orbit around Venus, effectively blocking or reflecting some of the Sun's rays. These solar shades would need to be large enough to reduce the amount of sunlight hitting Venus's atmosphere, gradually lowering the temperature. By reducing the Sun's radiation, these shades could help bring surface temperatures down over time, providing an initial step in making Venus more manageable for human technology and life.

Several sun-shading designs have been proposed, with some suggesting an orbiting "sunshade" that could be placed at the L1 Lagrange point between Venus and the Sun. At this point, the shade would stay in a stable position, consistently blocking sunlight from reaching the planet. This shade could be composed of highly reflective materials or even a system of small, interconnected mirrors that deflect sunlight away from Venus. Another design envisions a mesh or grid-like structure that could partially filter sunlight, allowing only a reduced amount of solar energy to reach the planet. Although creating and deploying a shade of this magnitude would be a massive engineering feat, it represents one of the most feasible approaches for gradually cooling Venus.

Reducing solar radiation would also set off a chain reaction within Venus's atmosphere. As the temperature drops, carbon dioxide, which makes up more than 96% of Venus's atmosphere, could begin to condense and even freeze out under the right conditions. By cooling the planet to a point where CO_2 solidifies, Venus's atmosphere could be significantly reduced in density, weakening its greenhouse effect. Over time, this would allow the surface to cool further, potentially to levels that could sustain more Earth-like conditions. However, managing this process would require precise control and understanding of atmospheric chemistry to avoid unintended consequences.

Another challenge is what to do with the vast amounts of carbon dioxide once it condenses. One vision for long-term terraforming involves deploying robotic systems or automated machinery to capture, store, or even eject the excess CO_2 from the planet's atmosphere. Some propose that massive "carbon elevators" or railgun systems could launch condensed carbon into space, slowly reducing Venus's greenhouse gases over centuries. Alternatively, stored CO_2 could potentially be used as a resource elsewhere, perhaps transported to other parts of the solar system or converted into industrial material.

While these ideas remain speculative, they form the foundation of a bold vision that pushes the boundaries of science and technology. Terraforming Venus is an ultimate test of planetary engineering, where challenges like atmospheric alteration, radiation shielding, and chemical control collide with the limits of human imagination. The process would require extensive scientific and technical breakthroughs, not to mention immense patience and resources, but in the distant future, it might be possible to transform Venus into a second home for humanity, rebalancing its environment to be compatible with life.

This long-term vision of a habitable Venus stretches our understanding of what's achievable, yet it remains a fascinating possibility, one that inspires new ways of thinking about planetary science, climate control, and the future of human existence beyond Earth.

A key step in the ambitious project of terraforming Venus would be addressing its dense, CO_2-heavy atmosphere, which traps heat in a runaway greenhouse effect. Reducing the carbon dioxide in Venus's atmosphere could not only lower surface temperatures but also bring its atmospheric composition closer to something that could eventually support life. This process, however, requires overcoming colossal challenges—Venus's atmosphere contains more than 96% carbon dioxide, creating pressures and temperatures that would be impossible to sustain for life as we know it.

One theoretical approach to reducing CO_2 involves cooling the atmosphere enough to trigger the condensation and eventual freezing of carbon dioxide. By deploying large sun-shading devices in orbit or at strategic points between Venus and the Sun, we could reduce the sunlight reaching the planet, allowing temperatures to fall significantly. As Venus cools, CO_2 could begin to solidify and freeze out of the atmosphere at around -78°C, given the right conditions. The idea is to gradually shift Venus's atmospheric temperature to the point where carbon dioxide becomes a solid, effectively "removing" it from the atmosphere and potentially creating solid CO_2 deposits on the surface or capturing it in some other way.

Once CO_2 begins to condense, several potential methods could be used to collect and dispose of it. One idea is to establish automated robotic systems on the surface that would gather and compress CO_2 into storage tanks or containment units, effectively removing it from the active atmosphere. Another, more speculative concept envisions mass-driver systems or railguns that could launch blocks of frozen CO_2 into space, gradually reducing the planet's atmospheric density by jettisoning material. If accomplished on a large enough scale, this removal could significantly alter Venus's climate over hundreds or thousands of years, providing the foundation for further atmospheric modifications that would bring it closer to Earth's conditions.

The excess CO_2 removed from Venus could, intriguingly, become a resource for transforming other planets. Mars, in contrast to Venus, has a thin atmosphere that lacks the density and greenhouse effect needed to trap heat effectively. This has led scientists to imagine a scenario where Venus and Mars could co-evolve, supporting each other in an extraordinary planetary transformation project. By transferring some of Venus's excess CO_2 to Mars, we could increase the Martian atmosphere's density, creating a greenhouse effect that would warm the planet. Over time, this influx of CO_2 could help Mars retain heat more efficiently, possibly warming the planet to a level where liquid water could exist on the surface—a critical step toward making Mars habitable.

Transporting CO_2 from Venus to Mars would be a grand-scale engineering feat, requiring interplanetary transportation systems capable of moving vast quantities of material between worlds. While speculative, some proposed concepts involve creating CO_2 "freight" capsules that could be launched in regular intervals, essentially acting as atmosphere transfers between planets. The transported CO_2 would contribute to building a more substantial Martian atmosphere, while

Venus's cooling effect would simultaneously progress as it loses its excess carbon dioxide.

This dual transformation, while deeply speculative, offers a vision of two planets evolving together, each benefiting from the other's unique challenges. Mars could gain a thicker atmosphere, warmer climate, and potentially accessible surface water, inching it closer to an environment that could support human life. Venus, in turn, would shed its thick, oppressive atmosphere, cooling to temperatures that would open the door for additional atmospheric modification and, eventually, surface exploration. In this vision, humanity's technological ingenuity would shape not just one planet but two, creating a solar system where both Venus and Mars evolve in tandem, supporting a future where life might thrive beyond Earth.

While still in the realm of imagination, this vision invites us to consider the possibilities of planetary engineering on a scale previously unimaginable.

Transforming Venus and Mars into habitable worlds wouldn't just be a triumph of science; it would be a profound reshaping of our place in the cosmos, creating opportunities for humanity to expand and thrive in ways that reflect the vast potential of our technology and vision.

Chapter 7: The Current Roadmap — Upcoming Venus Missions and Goals

NASA's renewed interest in Venus has brought forth two ambitious missions—DAVINCI+ and VERITAS—designed to unlock the mysteries of our neighboring planet. These missions mark a critical return to Venus after decades of focus on Mars and other planetary bodies, aiming to delve deep into Venus's atmosphere and geology to better understand why this "Earth twin" evolved so differently from our own planet. Each mission is tailored to tackle distinct aspects of Venusian science, combining advanced technology with a fresh determination to explore the secrets hidden beneath Venus's clouds.

The DAVINCI+ mission, short for "Deep Atmosphere Venus Investigation of Noble gases, Chemistry, and Imaging Plus," is set to investigate Venus's atmosphere with unprecedented detail. The mission's primary objective is to uncover clues about Venus's atmospheric composition and evolution, potentially revealing whether the planet once had an ocean or more Earth-like conditions. DAVINCI+ will deploy a descent sphere that will plunge through Venus's dense clouds, capturing precise data on noble gases and other atmospheric components as it makes its way down toward the surface. By analyzing elements like helium, neon, and argon, the mission seeks to uncover traces of Venus's atmospheric history, which can provide insights into how its climate spiraled into its current extreme state.

The descent sphere will also measure Venus's atmospheric chemistry and gather high-resolution images during its descent, offering a rare glimpse of the planet's rugged terrain and cloud structures. This data could help scientists determine if Venus ever had a habitable phase, giving us valuable information about the conditions necessary for sustaining life. If evidence suggests that Venus once had liquid water, it could reshape our understanding of planetary habitability and the potential for Earth-like worlds elsewhere in the galaxy.

VERITAS, the second mission, is focused on mapping Venus's surface in high detail to uncover its geologic history. Equipped with synthetic aperture radar, VERITAS will orbit Venus and use radar imaging to penetrate the thick clouds that conceal the planet's surface. This mission will provide topographic maps, revealing the structure and composition of Venus's crust. By studying these geological features, VERITAS aims to answer fundamental questions about Venus's tectonic activity, volcanic processes, and the reasons behind its geological divergence from Earth.

In particular, VERITAS will investigate whether Venus has any form of plate tectonics or ongoing volcanic activity, phenomena that are vital to understanding the planet's evolution. This information could reveal if Venus is still geologically active, offering clues about the processes that have shaped its landscape and whether it could still experience volcanic eruptions today. Together, DAVINCI+ and VERITAS represent a strategic approach to Venus exploration, with DAVINCI+ focusing on atmospheric analysis and VERITAS on geological mapping. These missions will work in tandem to paint a comprehensive picture of Venus, illuminating its past, present, and possible future as a planet that once may have been more like Earth than we ever imagined.

The VERITAS mission, short for "Venus Emissivity, Radio Science, InSAR, Topography, and Spectroscopy," is a groundbreaking effort to unravel Venus's geological secrets. By mapping the planet's surface with high-resolution radar, VERITAS aims to penetrate the dense clouds that obscure Venus, allowing scientists to create detailed topographic maps and gain insight into the planet's complex geology. This mission is designed to answer fundamental questions about Venus's tectonic structure, volcanic activity, and how its surface evolved in such a radically different direction from Earth's.

One of the primary goals of VERITAS is to assess whether Venus has any active tectonic or volcanic processes. Unlike Earth, Venus's surface lacks clear evidence of tectonic plate movement, which is essential for dissipating internal heat and shaping continents. However, VERITAS's radar imaging and interferometric synthetic aperture radar (InSAR) capabilities will enable scientists to look for subtle deformations in Venus's surface that might indicate active geologic processes. If VERITAS finds evidence of tectonic activity or recent volcanic flows, it could reshape our understanding of Venus's geological evolution and suggest that the planet may still be geologically active. These insights could reveal if Venus has been shaped by catastrophic volcanic episodes or if there's ongoing heat release, offering a new perspective on its tectonic makeup.

In addition to mapping topography, VERITAS will use spectroscopy to analyze surface composition, seeking clues about the mineralogy and volcanic history of Venus's crust. By identifying specific rock types, the mission can infer the conditions under which these rocks formed and assess whether there were once environments that might have been suitable for liquid water. The presence of certain minerals might even indicate ancient water activity, pushing the boundaries of what we know about Venus's capacity for habitability in the distant past.

The anticipated breakthroughs from DAVINCI+ and VERITAS could significantly impact the debate over Venus's suitability for exploration and potential colonization. By providing clear data on Venus's atmospheric and geological history, these missions may reveal if Venus once had more Earth-like conditions and if it has experienced water-related processes. If DAVINCI+ confirms that Venus's atmosphere contained signs of ancient oceans or stable climates, it would suggest that

Venus was once a more temperate world, raising questions about the longevity of such conditions and the factors that led to its dramatic transformation.

On the other hand, VERITAS's findings on tectonic and volcanic activity could reveal whether Venus remains geologically dynamic, which would have significant implications for human activity in its atmosphere. Active geology could mean that Venus continues to undergo transformations, influencing its atmospheric composition over time. This knowledge would be crucial for future atmospheric habitats, as understanding Venus's geological processes could help in predicting long-term atmospheric stability for floating colonies.

Together, these missions offer the promise of a more nuanced understanding of Venus, shedding light on whether this volatile planet could support human exploration and even habitation in its upper atmosphere. By answering essential questions about Venus's past and present, DAVINCI+ and

VERITAS have the potential to redefine our perspective on Venus, not just as a scientific curiosity but as a possible destination for human ingenuity, exploration, and survival beyond Earth.

Chapter 8: The Role of Private Space Exploration in Venus Colonization

SpaceX has ignited global interest in Mars through its ambitious plans to make humanity a multiplanetary species. Driven by the vision of Elon Musk, the company's Mars-centric goals aim to establish a sustainable human presence on the Red Planet. Central to this vision is the development of the Starship, a fully reusable spacecraft designed to carry large crews and heavy cargo across interplanetary distances. By significantly reducing launch costs and increasing payload capacity, SpaceX hopes to make space travel more accessible, ultimately enabling Mars colonization with regular supply and passenger flights. The goal is to establish Mars as a "backup" for Earth, a place where human civilization can thrive and expand. This approach, which leverages private funding, innovation, and rapid iteration, has reshaped the field of space exploration and is redefining what private spaceflight can accomplish.

The question of whether SpaceX could apply this vision to Venus is intriguing, though the challenges would be markedly different. Unlike Mars, where a stable surface and predictable environment allow for traditional landers and rovers, Venus's intense heat, atmospheric pressure, and corrosive clouds demand unique technologies. However, SpaceX's strengths in innovation, large-scale manufacturing, and cost-effective launches position it as a potential pioneer for Venus exploration. Starship, with its heavy-lift capacity, could deliver large payloads, atmospheric probes, or floating habitats to Venusian orbit, enabling new types of scientific missions that require robust equipment and specialized materials.

One feasible role for SpaceX in Venus exploration could involve deploying high-altitude atmospheric platforms or probes capable of floating in the planet's temperate cloud layer. Given Starship's capacity to transport substantial equipment, SpaceX could support missions involving solar-powered drones or balloons that collect atmospheric data and send back information. This approach aligns well with SpaceX's mission of making space exploration economical; by delivering large payloads of probes or equipment to Venus's upper atmosphere, the company could facilitate long-duration science missions at a fraction of the typical cost.

Moreover, SpaceX's drive toward rapid prototyping and modular systems could accelerate the development of specialized Venusian equipment. While Venus presents daunting environmental challenges, the possibility of a floating colony in the "Goldilocks Zone" of its atmosphere could be explored through small-scale experiments. SpaceX could collaborate with scientific organizations to test acid-resistant materials, autonomous data-gathering devices, or even scaled-down versions of floating habitats. With the company's agile approach to engineering and relentless focus on cost reduction, SpaceX could help pave the way for future atmospheric missions to Venus.

While Venus is not currently part of SpaceX's primary vision, the company's Mars-focused goals may naturally lead to an interest in other worlds over time. If Mars colonization proves successful, and SpaceX achieves its ambition of routine interplanetary travel, then expanding to Venus—albeit in its upper atmosphere—could be a logical progression. While Mars offers a more direct path for colonization, Venus presents a valuable opportunity to test new technologies in extreme conditions, an endeavor that could deepen humanity's capabilities in planetary exploration. SpaceX, with its bold vision and groundbreaking technology, has the potential to extend humanity's reach not only to Mars but to the mysterious skies of Venus, showcasing the transformative power of private spaceflight in venturing into new frontiers.

As humanity sets its sights on Mars as the first destination for interplanetary colonization, the experience, technologies, and infrastructures developed for Mars missions could play a crucial role in opening the door to future exploration and potential settlement of Venus. Establishing a sustained presence on Mars will require overcoming extreme challenges—creating life-support systems, harnessing in-situ resources, developing renewable energy sources, and building habitats resilient to harsh conditions. These achievements, though aimed at Mars, would provide a wealth of knowledge and practical solutions that could be adapted to other planets, including Venus.

One of the primary challenges on Mars is generating and recycling vital resources such as water, oxygen, and fuel. Techniques for extracting water from Martian ice and producing oxygen from the carbon dioxide-heavy atmosphere, for example, would be invaluable for a future mission to Venus. In Venus's upper atmosphere, where CO_2 is abundant, similar technologies could be used to generate breathable air and fuel for Venusian habitats. Mars missions are also likely to pioneer closed-loop life-support systems and waste recycling methods that could be directly applied in Venus's atmosphere, where access to external resources would be equally limited.

Additionally, Mars colonization will likely drive advancements in autonomous robotics and remote-controlled systems, which are essential for Venus missions. Venus's extreme surface conditions make human exploration unfeasible, but high-altitude drones, robotic probes, and automated systems could operate in the planet's temperate atmospheric layers. These autonomous technologies, developed and refined through years of Mars missions, could be repurposed to navigate Venus's dense atmosphere and corrosive conditions, enhancing humanity's reach beyond what traditional spacecraft alone could achieve.

Energy solutions developed for Mars could also support future Venus missions. Mars's distance from the Sun requires efficient solar panels, nuclear power options, and advanced energy storage systems to maintain consistent energy supplies in an environment where solar energy is less effective. These energy innovations, essential for sustaining Mars colonies, could prove equally beneficial in Venus's atmosphere, where solar power could play a critical role in maintaining floating habitats. Likewise, the technology for energy storage and distribution developed on Mars would be vital for powering operations in Venus's cloud layers, ensuring that habitats can operate sustainably above the planet's surface.

Perhaps one of the most significant contributions Mars exploration could offer to Venus missions is in the realm of habitat design and construction. Building resilient, modular, and possibly inflatable habitats for Mars's surface will provide insight into creating adaptable structures that could be modified to float in Venus's atmosphere. The knowledge gained from constructing habitats that can withstand Mars's dust storms, radiation, and freezing temperatures could guide the design of acid-resistant, lightweight habitats for Venus's more temperate "Goldilocks Zone." These floating habitats, built on the foundation of Mars's modular and expandable habitat technology, could potentially support long-term human habitation in Venus's upper atmosphere.

In many ways, Mars can be viewed as the proving ground for humanity's interplanetary future, with each innovation paving the way for expanding into new worlds. By addressing Mars's extreme challenges, humanity will develop the skills, resilience, and technologies that will make exploring Venus feasible. Mars colonization will not only demonstrate that life can thrive beyond Earth but also provide the tools and confidence needed to extend that life to other parts of the solar system, including the enigmatic clouds of Venus. With each step, humanity moves closer to realizing a future where interplanetary exploration is not limited to one world but forms a network of thriving outposts, each supporting the other in a pioneering vision of life among the stars.

Chapter 9: The Future of Multiplanetary Civilization

Venus occupies a unique position in humanity's long-term vision of becoming a multiplanetary species. While Mars often takes center stage as the most feasible destination for colonization, Venus offers a different kind of opportunity—one that challenges our understanding of survival and adaptation beyond Earth. As a neighboring planet with striking similarities to Earth, yet radically different conditions, Venus provides an invaluable platform for scientific discovery, technological advancement, and the exploration of what it truly means to expand life beyond our home planet. Its extreme environment pushes us to innovate, think creatively, and test the very limits of human ingenuity. By incorporating Venus into humanity's interplanetary aspirations, we diversify our approach, creating pathways for exploration that are as varied as the planets themselves.

In the larger picture, Venus represents a crucial step toward achieving a multiplanetary future. While Mars offers a "second Earth" on solid ground, Venus challenges us to imagine a new kind of habitat: floating cities that exist in a stable layer of the planet's atmosphere, safe from its hostile surface. By developing technology to sustain life in such a unique environment, we broaden our capabilities, preparing for scenarios that may await us elsewhere in the universe. The expertise gained in creating acid-resistant habitats, harvesting resources from dense atmospheres, and maintaining buoyant structures could one day be applied to other planets, moons, or even rogue bodies with thick atmospheres, thereby expanding humanity's reach in the cosmos.

Venus also offers a critical contribution to humanity's resilience, serving as a potential "Plan B" that enhances our adaptability and long-term survival. By establishing colonies that can thrive in Venus's upper atmosphere, we create a lifeline in the face of unknown challenges—whether environmental, technological, or societal. Venusian colonies would not only provide an alternative refuge but also serve as a base for scientific research on planetary climates, atmospheric chemistry, and environmental stability. Understanding Venus's extreme greenhouse effect, for instance, could yield vital insights into Earth's climate, allowing us to better manage and perhaps mitigate the effects of climate change. Through this knowledge, we would not only make strides in Venus exploration but also gain valuable information that could protect life on Earth.

A Venusian colony would also expand our scientific and cultural horizons. Just as the International Space Station has fostered unprecedented collaboration and innovation, a presence in Venus's atmosphere could unite nations, scientists, and thinkers in the shared pursuit of exploration. By learning to coexist and thrive in such a challenging environment, humanity would develop new perspectives on cooperation, sustainability, and the possibilities of life in places far different from Earth. The scientific advances from studying Venus's volatile climate and geological features could spark innovations that reshape industries back on Earth, from energy production to environmental management, as we adapt and apply technologies initially developed to survive in Venus's skies.

Ultimately, Venus fits into the larger vision of a multiplanetary future by providing humanity with a new frontier—one that complements Mars, extends our reach, and strengthens our adaptability. In building the skills and technologies to sustain life in Venus's atmosphere, we would grow more resilient as a species, gaining the flexibility to explore, survive, and flourish in a variety of extraterrestrial environments. As Earth's closest planetary sibling, Venus holds a mirror up to both our aspirations and our limitations, inviting us to push forward, explore the unknown, and embrace the challenges of expanding life beyond Earth. In this way, Venus doesn't just serve as a "Plan B"; it becomes a vital component of our evolving story, adding richness, diversity, and resilience to humanity's journey among the stars.

Colonizing Venus could be a stepping stone in humanity's journey to reach beyond our solar system, as the innovations and skills developed for Venusian habitats would prepare us for the unknown environments awaiting us in distant star systems. Venus, with its dense atmosphere, extreme heat, and crushing pressure at the surface, challenges our ability to adapt, pushing us to develop technologies that enable survival in hostile and unconventional settings. As we master the art of creating life-supporting environments in Venus's clouds, we gain essential experience in building sustainable habitats that could one day float in the atmospheres of exoplanets, orbit around other stars, or endure extreme climates in ways not limited to terrestrial assumptions.

The process of creating floating colonies above Venus's surface would provide a wealth of knowledge applicable to exoplanetary exploration. Techniques for autonomous energy generation, atmospheric resource harvesting, and closed-loop ecosystems would all be honed on Venus, equipping humanity with the tools to establish outposts in similarly challenging locations across the galaxy. Many of the exoplanets discovered in recent years have dense atmospheres, varied gravity, or intense climates—worlds that could resemble Venus in many respects. By learning to thrive in Venus's skies, we gain the flexibility to explore and survive on these far-off planets, prepared for environments that might otherwise be considered inhospitable.

Venus could also be the proving ground for advanced propulsion and space travel systems needed for interstellar exploration. The ability to operate autonomously and generate life-sustaining resources on Venus would be crucial for missions that span decades or centuries, enabling deep-space travel that is self-sufficient and sustainable. Once we can reliably create and maintain life-support systems in Venus's atmosphere, these same technologies could be adapted for long-duration journeys to nearby stars, where astronauts or explorers may have to rely on onboard resource recycling for years. In this sense, Venus doesn't just offer an opportunity for habitation within our solar system—it could serve as a training ground for the resilience and adaptability required for interstellar exploration.

As we consider the future of humanity's journey beyond Earth, the potential to expand into places as varied as Mars, Venus, and even interstellar space demonstrates the limitless scope of human ingenuity. Colonizing Venus invites us to confront assumptions about where and how life can exist, challenging us to push beyond familiar boundaries and conventional thinking. Human curiosity has driven exploration for millennia, leading us to scale mountains, cross oceans, and, more recently, venture into the cosmos. Each new frontier reminds us of the remarkable capacity within us to transform, to innovate, and to grow in ways that once seemed unimaginable.

Exploring Venus—and ultimately other star systems—symbolizes a profound shift in our understanding of what it means to be human. It's not simply about survival or the practical expansion of civilization; it's about embracing the transformative power of exploration, about reimagining our place in the universe. In the face of challenges, the drive to learn and discover remains one of our greatest strengths, shaping a future where humanity thrives across worlds and dimensions we've yet to comprehend. By continuing to push forward, to challenge our assumptions, and to dare what seems impossible, we illuminate the vast potential of our species—a boundless curiosity that extends to the stars, seeking to understand not just the universe, but our own capacity for wonder and resilience.

Conclusion

Venus, once dismissed as a fiery, inhospitable world, emerges through closer examination as a remarkable candidate for exploration and possibly even habitation. Throughout this journey, we've uncovered that Venus is not simply an exotic neighbor; it is a scientifically rich, strategically valuable world that challenges our assumptions about where life can thrive and how far human ingenuity can reach. While Mars may hold the promise of a second Earth on solid ground, Venus invites us to imagine a future in the skies, where floating cities could transform its atmosphere into a realm of research, exploration, and sustainable living. By focusing on the temperate layers of Venus's atmosphere, we're not merely pursuing curiosity—we're taking a strategic leap forward, developing technologies and ideas that could expand our options for life beyond Earth.

Venus's thick atmosphere, far from being an obstacle, offers a unique advantage: it provides a stable environment at around 50 kilometers above the surface where Earth-like pressures and temperatures open a new frontier for atmospheric habitats. Here, humans could live in cities that hover gracefully above the planet's hostile surface, protected by advanced, acid-resistant materials and sustainable life-support systems. These floating cities would serve not only as homes but as centers of scientific discovery, observing the complex chemical dynamics of Venus's atmosphere, studying its volcanic landscape from above, and perhaps uncovering clues about how planets evolve and how climates can shift so dramatically. In this vision of the future, Venus becomes a beacon of human resilience and adaptability—a place where innovation flourishes and expands our understanding of what it means to be part of a multiplanetary civilization.

The pursuit of Venus is not just about claiming a new frontier; it's about fostering a spirit of exploration and investing in knowledge that could one day be crucial for the survival and progress of humanity. In exploring Venus, we strengthen our ability to innovate in extreme environments, building tools and techniques that may one day carry us to other worlds. The breakthroughs achieved on Venus—whether in climate science, atmospheric chemistry, or advanced engineering—will resonate beyond its atmosphere, impacting life on Earth and laying the groundwork for humanity's expansion into deeper parts of the cosmos.

As we close this exploration of Venus, let it serve as a call to action: a call for curiosity, for investment in science and technology, and for a renewed commitment to discovery. Venus deserves a second look, not only for what it can teach us but for the inspiration it ignites to reach further, to imagine more boldly, and to push the boundaries of what is possible. In choosing Venus, we choose to embrace a future where humanity's horizons are ever-expanding, a future that calls us to look beyond the familiar and to pursue the unknown with courage and determination. With Venus, we have the opportunity to make history—not by conquering a new land, but by redefining our place among the stars.

www.ingramcontent.com/pod-product-compliance
Lightning Source LLC
Chambersburg PA
CBHW070300220526
45465CB00004B/1684